Solution Publishing Co.
26 Welton St., Milford, CT. 06460

FOCUSING MATTER

Everything is moving: motion, existence, and time, are one and the same thing. Solid mass is the focusing (implosive movement) of spheres of infinite sizes; these spheres form each other.   A table maintains its shape because the spheres that are forming it continue to have the same pattern of motion toward its inner spaces. Motion can be altered through interference. Solid mass is the obstruction of one or more types of motion in favor of other types.   A particle (sphere) is a close knit group of particles (a packet); each packet is made up of smaller packets to infinity.

A planet is formed by both an imploding and exploding motion. Planets are formed at the border of three or more solar systems. The number of solar systems meeting at the same point determines the dominant polarity of the planet (more on that later).   The head-on collisions of expulsed particles, from these solar systems eventually form a planet destined to become a sun; or as solar systems shift positions the planet falls out of alignment and hurtles through space (like our own planet).

As gravity particles move through space they meet less resistance and have increasingly fewer collisions; Their momentum increases until they collide with similar sized particles, or tight knit net of particles.

The planets in our own solar system were not formed here but knocked here by shifting solar systems. When a planet is knowcked out of the point of collision of small particles, the small particles inside the planet sphere continue their journey toward the center of the sphere, through the spaces between more dense packets, leaving relatively empty space behind them (space with fewer collisions increasing frequency of motion); This allows gravity particles to continue moving into the planet. As they move toward the center of the planet they meet each other to form a shell; like Laurel and Hardy trying to fit through a door at the same time. Small constituents of the gravity packets (in the shell) continue on toward inner space. Incoming gravity spheres collide with the spheres making up the shell, this causes the striking sphere to reverse (see p. ¶ Smaller faster sphere striking motionless or slower sphere) polarity and send its ultra small constituents in an outward (explosive) direction (negative polarity). When on-coming particles are moving at a great enough

4

velocity the polarity change can be great enough to force the spheres to move violently outward, remaining in tact as they pass out of the planet or atom; this is Atomic Radiation such as Alfa Waves.

The original Lattice (Matrix) (Stencil) of the shell toward the center of the planet still has the same pattern today; With the exception of changes caused by asteroids having entered the earth forming magma which retains much of its original polarity and shape; Also causing the movement of continental plates. (polarity will be defined later on).

As particles are passing out of a sphere they pick up volume due to many additive collisions. As these particles head for open space there are fewer collisions, (implosion frequency slows down), so they are losing volume faster than they are gaining volume; The size of the spheres continues to decrease as it moves farther from other similar sized spheres.

These long traveled small extra fast particles will eventually collide head-on into a point of focus, those missing the focus will still collide head-on forming a lattice of motionless particles with double the volume. On second impact by a small travelled particle the new motionless particle is shattered; into how many peices depends on the shape of the matrix. Further collisions will somewhat evenly divide the direction of flow.

The closer particles come to a planet (focus point) the larger they become.

CHAPTER II

I stands for infinite.

<u>proof</u>

I ÷ 1 = 0	1 × 0 = I	
0 ÷ 1 = 0	1 × 0 = 0	{I=0}
I ÷ I = 1 {I÷I=I}	I × 1 = I	
I ÷ I = I	1 × 1 = 1	{I=1God}

This equation proving infinity equals zero is incorrect yet is correct; this paradox is due to the fact that our existence and math are products, of the mind of one being. But remember the fractions of one are infinite 1 = I; {I÷I=X} X being any sum except zero.

ORBITAL STABILITY OF SPHERES

The above mathematical paradox proves that matter, motion, and time are one and the same thing.   It also proves that there can be no explanation for why thigs exist and move.   Taken that things do move! and that this movement is an infinite series of reflections, then it can be assumed that a ball can bounce, and that gravitational forces can exist.

Discription of Gravity and Orbital Stability of Planetary, and Atomic Spheres: When expulsed particles from the planets collide, they collide at outward angles deflecting outward away from the direction

7

of both planets, and fall into the orbit of, and are parted by, the less dense planet (more positive sphere) this creates a partial vacuum (removes larger gravity spheres) between the two planets creating less resistance to incoming (attractive) movement.

As the planets move together the expulsed particles from the planets are more concentrated, they are also larger and slower moving; fewer particle types being able to pass through the planets, and the matrix of implosive particles entering the planets. The expulsed particles moving out from in between the planets are moving much slower; Smaller expulsive particles are being expulsed from accumulating numbers of the original expulsed particles; these particles are colliding forming bispheres and parallel lines of resistance between the attracted planets. The result is a perfect non-disintegrating orbit (perfect gravitational equilibrium).

Planets circle when in orbit since the orbiting planet was not in a perfect line of collision on its approach or collision with the planet it orbits.

Orbiting excites greater negative force (expulsivity) explained later in this text.

People and other objects on the surface of a sphere are attracted in the same way as another similar sized sphere, but are not forced into orbit (repulsed), due to a lack of particles expulsed from the grounded body. Greater pressure (collision frequency) builds up between one sphere and another and between the sphere and atmosphere above the human body in question on the sphere's surface.

If an object underground, water, or atmosphere; allows implosive gravity to pass through it more readily than the surrounding media, then the expulsive force of the surrounding material will strike the object with greater force, exciting its expulsive force; The expulsed particles increase collision density with expulsed particles from the core of the sphere, but not with smaller incoming particles since they fit through the lattice of larger negatively charged spheres (particles), causing it to rise through the medium (be expulsed further from the core of the sphere.

CHAPTER III

When two spheres of the same size and
density moving about the same speed collide
head-on: The dominant polarity (Dominant
movement of all of the constituents of the
sphere and within the line of flight)-   As
the two spheres collide the first points of
impact are pinched inward; The constituents
affected by the impact as they turn inward
their polarity is turned inward as well; the
opposing sides form a new core which the
back outer surfaces are pulled toward, rotating their polarity in perfect counter
symmetry so that force is met with equal
force, creating one perfect sphere from two.
Its momentum stalled until another uneven
force creates greater impacts (condensed
frequency pressure, or faster frequency)
(greater density force by smaller particles)
or (less density greater velocity force by
larger particles).

Taken that two spheres of similar density
and equal size moving at different veloci-
ties, collide head-on: Force against the
slower sphere is not even therefore the outward bending polarity of the hind particals
of the slower sphere move these particles

11

outward over the outer surface of the faster
sphere pulling and expanding the core of the
slower sphere creating a Collition Intensity
Void; which pulls (allowes to pass) the center of the faster sphere through the slower
sphere creating one particle which is continually turning inside-out and alternating
polarity (see fig.2)

If the sphere with alternating polarity
was forced to move at a right angle to both
its poles and then was struck by a faster
sphere (head-on):   the   alternating polar
sphere would   come   into   balance (Polar E-
quilibrium);   Then would have a greater at-
traction for smaller particles and greater
expulsive force toward larger particles.

When   two   spheres   of   different   size,
moving   at   the   same   speed,   having   equal
density collide head-on: Has basically the
same sequence as for two spheres of the same
size   same   speed   colliding;   only   that   an
extra core is created toward the surface of
the   larger   sphere,   and   a   mound   created
projecting from the surface of the larger
sphere.

In a similar collision taken that the
larger sphere is less solid (more liquid)
than the smaller;   The smaller sphere would
be swallowed by the larger and then would
orbit the core of the larger sphere (even if

12

the smaller sphere melted it would maintain its polarity as a sphere, and maintain its orbit in the larger sphere).

In a similar collision taken that both spheres have loose density, the outer perimeter of the larger sphere surounding the impact zone would continue moving forward; the polarity of which bends around the smaller sphere, focusing above the smaller sphere; this bending causes an expansion (growth) of colliding expulsive particles, above the smaller sphere, pressing the smaller sphere toward the core of the larger sphere; becoming a part of that core at the shell level having equal density.

The head-on collision of two equally dense particles, one larger and one smaller. (Dense as in strong inward and outward movement, to and from the constituents of the packet). Taken that the smaller sphere is moving at a greater velocity: Due to the greater velocity of the smaller particle, more of its smaller constituents meet less resistance and continue moving forward, this causes greater negativity in the larger slower sphere, causing constituents of the smaller faster sphere to reflect [see fig.3] causes it to be spread out creating a void in its core: which attracts the pinched inward particles at the area of impact, turning the smaller sphere inside out; its con-

stituents moving in about-face, spout out of
the back of the sphere, spraying its constituents outward over the larger sphere; their
polarity reversed contributing to the larger
spheres own polarity, accelerating its momentum initially; then over all negativity
(expulsivity) as the smaller spheres remains
spread out over the larger sphere; A crater
is left, (see fig.3).

When a faster smaller sphere overtakes a
slower larger sphere, both moving away from
the same direction; Same line of flight: The
particles of the larger sphere at the area
of impact have their polarity turned outward
instead of inward, peeling the exosphere
(outer surface) out away from the core, this
separated material expands dividing into a
number of small spheres; The core is then
left off-center at the back surface of what
is left of the large sphere; (while this is
occurring the smaller sphere has turned
inside out, reversing polarity and direction
of flight). The core being set outside the
center of the sphere causes it to orbit the
center, causing the sphere to spin; The
smaller sphere having reversed polarity,
reaches cymmetrical equilibrium with the
larger sphere's core, polar force meeting
equal polar force. As the sphere spins,
the side spinning against (in opposition to)

it's own wake forces the sphere to move to-
word the side moving with (in favor of) the
oncoming particles in the wake of the sphere
(see Chap.V); Forming a spiral such as
seen in the Target Bubble Chamber of an
Acceletron. (fig.4).

Taken that two spheres of similar density
and Equal Size, moving at different vel-
ocities, in the same direction and line of
flight; the faster sphere overtaking the
slower sphere. The particles in the wake of
the first (slower) sphere have their polarity reversed when struck by the faster sphere
creating a build up of negative force between the to spheres which allows both
spheres to turn inside out, but the greater
negative force of the faster sphere does not
allow the slower sphere to reverse polarity;
causing them to push apart and fly off in
opposite directions. (see fig. 7)

fig.1

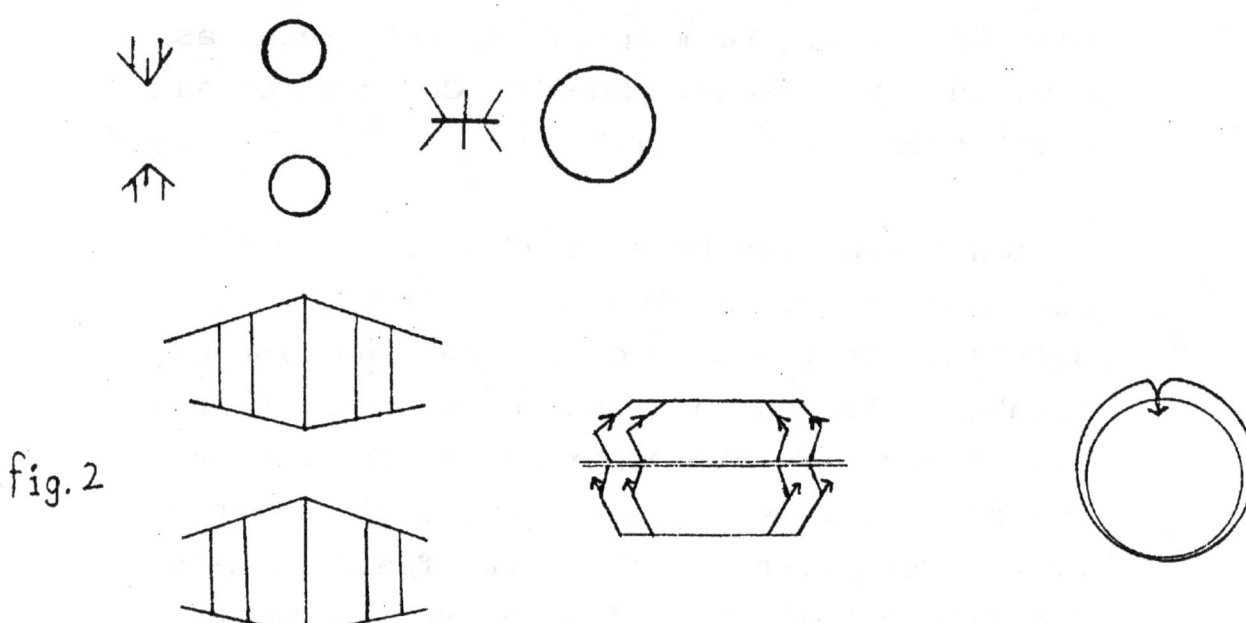

fig.2

fig.3

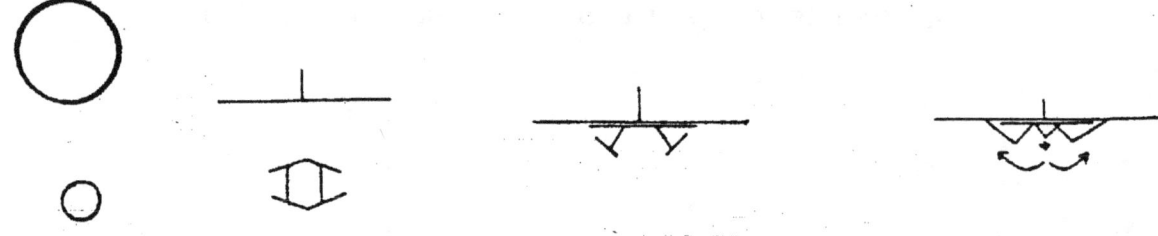

arthur P. Yarado

REFLECTION--   As a ball is thrown through
the air all of its constituents, and all of
the constituents of existence are put into
motion.   When the ball strikes a wall even
if it is a ball of stone it is reflected by
the excitation of expulsive movement of constituents (parts) of the wall as they are
struck by ultra small particles sandwiched
between the core of the ball and the cores
of the wall. If the ball strikes at an
angle the incoming wake (implosive particles
following the ball) will only be able to
move the ball on the side of the ball which
is furthest from the wall (has the least
sandwiched pressure between the wall and the
ball) in relativity to the angle of impact;
This causes the ball to roll.   Since particles inside the ball are moving in the
same direction (in relativity to the core)
they make room for particles entering the
ball at the same new angle (in respect to
the wall) when this angle of expulsion
collides with the incoming particles in the
wake so that biospheres are formed (two
combining to form one) so that they form a
wave pattern to the wake, to the wall, then

back to the ball again, only then does the
ball stop spinning and fly off the wall in
the opposite direction from its original
dominant internal polarity.   (DOMINENT
POLARITY- direction of dominant flow of constituent particles).   (see fig.8).

CHAPTER V

REFLECTIONS INTO A SPIRAL

HOW A TOP KEEPS BALANCE-- When a top is spun implosive gravity particles which prior to the spinning of the top where passing into the top through a lattice pattern (stencil) of particles in stable orbits and outward moving shells. Gravity particles move in a lattice formation themselves; through the lattice shells of expulsed particles. As the top is spun the solid parts of the lattice strikes or bends the incoming parti-cles. The incoming particles then squeeze (wave collide with each other) as the path of the interfered with particles bend in the direction their original entrance holes are moving, this creates a chain reaction which bends (changes the path formation) of all incoming particles (of earth gravity size). This formation is that of spiralling lines of interference (whirl pool, or tornado type reaction). The spiral movement of incoming particles is continued inside the Top due to the expulsed particles which continue to move in a straight line even though the Top is rotating; therefore creating spiral re-sistance lines. To observe this draw a circle on a paper then move the pencil in a compass in a straight line as you rotate the paper.   The resistance lines even though they alter the path of particles; those particles maintain the same pattern of distribution therefore the same matrix pattern.

Since the Top only rotates in a two-dimensional plane, expulsive movement is only excited in the two dimensional plane that it rotates in.   Expulsive particles from the earth's core leave the planet in a relatively straight line those particles entering the planet are also moving in a straight line.   The reflections bend into a spiral and having greater frequency of collision (pressure, volume) in the plane that it rotates; If the Top tilted the particles entering and leaving the Top would collide in head-on Bi-sphere collisions with particles entering and leaving the earth; the angle reflection is such that particle density builds up on the side facing into particle flow; this balances the Top horizontal to the plane of the earth. Collisions perpendicular to the plane of the Top's rotation are then equal on all sides.

The Top when spinning creates a greater number of collisions therefore greater mass but since the motion is two-dimentional, so is the mass.   The Top has more orbital stability in the horizontal plane that it is spinning in, than above or below; therefore it has more resistance to horizontal movement than when not spinning; Spiral lines of resistance force incoming particles from above and outgoing particles from below, outward horizontally; explaining greater resistance to (kinetic) horizontal motion. To a lesser extent downward and upward motion is also resisted.

A Tornado just like a Top resists movement. The tornado tends to stay in place as the earth rotates beneath it; due to irregularities in the topography, and diverse expulsive gravity lines of polarity in the mantle (ledge) covering the earth (due to folds and movement of magma beneath). A compromise is struck between the movement of expulsed gravity particles (direction of) from the crust (mantle) and the movement of the earth (sphere) and the rims of hills, solid buildings, and trees. The winds have no bearing on the Tornado's movement except to give birth to the tornado and then to destroy it.

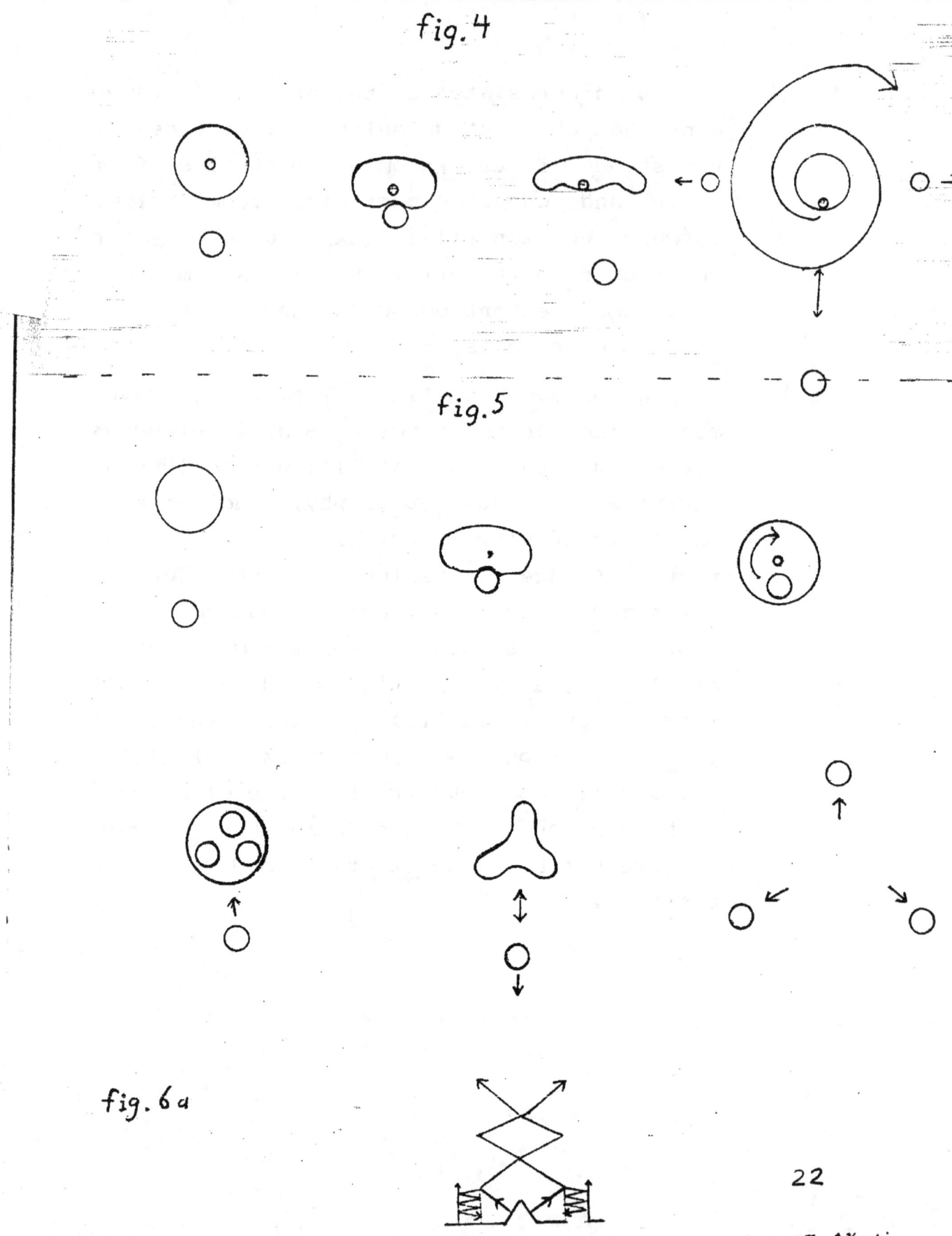

fig. 4

fig. 5

fig. 6 a

22

CHAPTER VI

RESONANCE & WAVES

How a glass is shattered by sound: Different elements are spheres of different sizes, densities, and polar configuration, therefore each element is spaced differently so the time it takes Element-Spheres to reflect off of other Element-Spheres after being impacted themselves, depends both on how hard they are impacted and how far apart they are. If air waves (sound) impacts them while they are on the rebound, then the movement of the Elementspheres is restrained while movement of the constituent particles (inertia particles) is increased. When the air waves strike slower than the crystalline waves (Element-Sphere Collisions), so that the surface of the glass is struck only after the reflected Element-Spheres have reached the end of their elasticity (their induced negativity is reversed by head-on collisions), then the gas atom wave (sound wave) strikes adding each time to the reversal of glass atom polarity; each time the glass stretches (bends) farther, until it collapses (shatters).

fig. 6B

SLOWER

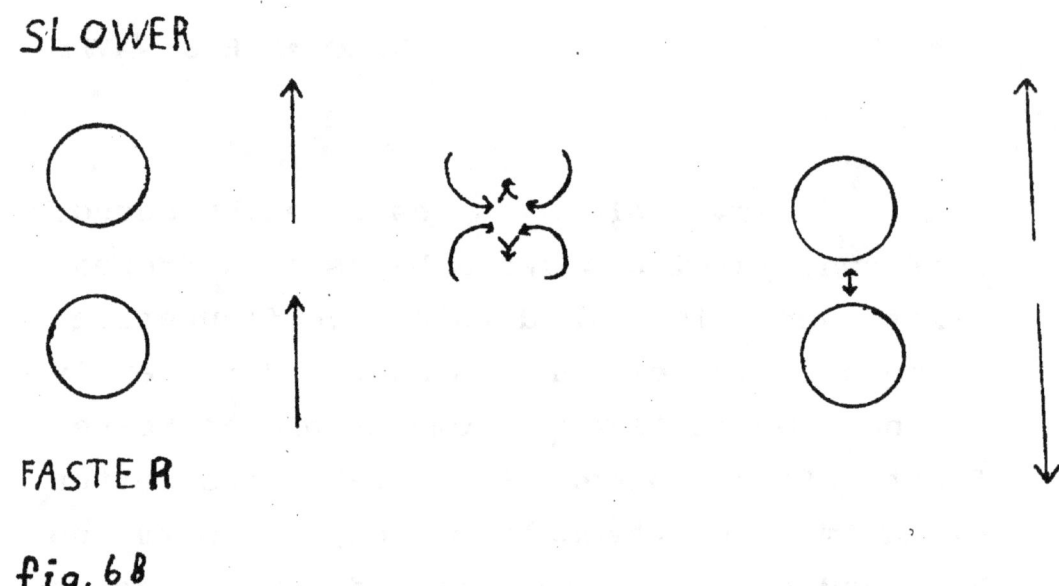

FASTER

fig. 6B

fig. 7 Reflection

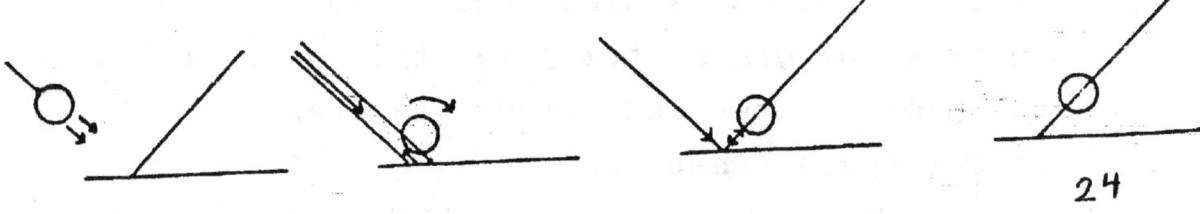

24

CHAPTER VII

GENERATION OF ELECTRICITY

Why electricity moves through a wire passed between opposite poles of a magnet: Metal atoms are made from collisions where smaller spheres enter larger spheres and then orbit the larger sphere's core. This creates greater negative force (expulsive movement) in the plane of the orbiting smaller sphere. This causes the atoms to align themselves alternately so the planes of orbits in neighboring atoms are not in the same plane.

All of the orbitting bodies set themselves as far apart from each other as possible; therefore every other atom rotates in the same direction at the same speed. The atoms with an electron orbital (rotating) plane horizontal to the plane that the magnet particles (from the generator magnet) are flowing, have their electron particles (formed from expulsive particles) shoved in the direction of magnetic particle flow; This causes the electron packets to accumulate on the opposite side of the Element-sphere (atom); Causing the atoms to be heavy (have more mass) on one side, as a result the electron orbits in question are forced on their side when the wire is moved up or down through the magnetic flow of a magnet.

If the wire atoms have orbits perpendicular (blocking) the path of the magnetic flow no single electron packet is formed; The electron charge of these orbits is increased; An electron packet is only formed when the wire is lowered or raised. In a slingshot reaction the electron packet on half a turn flies out of orbit. Each atom is a gyroscope resisting rotational motion. All Copper, Aluminum, and precious metals have electron orbits rotating counterclockwise when facing north, regardless of the direction the wire is pointed in. When the wire is lowered electrons fly off to the right when raised fly off to the left. Vertically orbited atoms inbetween the generating atoms have their electron (magnetic) particles forced up when the wire is lowered giving the freed electrons an open (underhand) projectory; and forced down when raised giving a clear overhand projectory to the left.

Magnetic particles maintain the flight formation they had inside the magnet after leaving the magnet so there are spaces between each single file row of magnet particles; Allowing the wire atoms to alternately tip then right themselves, then tip.. as they are moved through the magnetic field.

CHAPTER VIII

RESISTANCE TO THERMAL MOTION

<u>HOW A THERMOS WORKS</u>--   In a partial vacuum particles tend to lose their constituents faster.   These particles form intense resistence lines (coherent through ways) between atoms; which resist the passing of heat particles through the thermos container

Page 28

CHAPTER VIIII

$$E = Mc^2$$

How does Einstein's theory $E = MC^2$: Energy = Matter × The Speed of Light Squared; stand up to the Unification Theory: Taking one single size of particle; assuming it is the size of a light particle; these particles making up a sphere, if the sphere expands to: itself squared, then the outer shell of particles will reach a maximum speed of the speed of light (at sea level) squared.

Page 30

CHAPTER X

TRANSPARENCY & LENSES

emitted from glass atoms reverse polarity after striking light spheres: offering no resistance to the flow of light from direct collisions; Electron packets formed at the borders of glass atoms have sufficient repulsive force to navigate light particles around them; This could cause light particles to collide on the opposite side of the more negative electron or atom; if they are only squeezed closer together magnetic waves are evolved from excitation of expulsivity; if they collide on the opposite side of an atom or electron and bispheres form, ultraviolet will become visible-violet, and infrared will become visible-red;

The stew of different colors would remain an even mix unless the atomic structure is such that light particles of different sizes are sorted by atomic gravity in different ways in their varying channels. Iceland Spar splits light evenly into two different directions doubling images. Some plastics can change ultra violet into visible light while keeping visible light from changing, making the plastic seem to glow; it can even increase visibility in a room.

How A Picture Is Focused: The atoms in glass give off particles which interfere with the flow of particles given off by light spheres (particles) as a result, inside the corridors of the glass, products of light particles are ferried to and obsorbed by the glass atoms; Outside the surface of the glass since there are no equal sided collisions between larger particles expelled from atoms and smaller particles expelled from light: less density (impedance) develops between the light and the glass; Causing light to turn in toward the densest areas of the glass (lens). This is demonstrated by the concave lens effect from a flat pain of glass (spreading of focussed light without difusement).

Near objects reflect more angles of light and more light particles per square inch of lens space, than distant objects. Light particles from close objects are turned less inward by a convex lens due to increased collisions in the dense more divergent angles of close subjects. Distant light has spread out therefore is less dense and more easily inflected.

To recreate a sharp image (picture) the lens must sort the angles of light: By allowing the particles (moving from a simular angle) to invert (condense) while still diverging (separating) from beams coming from different angles. Ricocheting light reaches the lens from all directions. Standing in one spot all points (origin) of light reaches you from different angles; When passing through a PIN HOLE the mixture of light particles spread out in different directions; Since there is no other light obscuring them dots of light the size of the pin hole recreate the scene upside down and reversed. A lens bends light more than its original angle, thereby widening the difference between focal planes. When near things are focused distant things are put so far past focus it blends in, increasing brightness but not obscurring the scene.

A diaphram (Iris) is placed between lenses so

that distant objects which get bent more inward
pass through the pupil while close objects are
still spread out and reach the second lens to a
lesser degree; If too much of the near light is
screened a light spot would form in the center of
the picture; The living eye copes with this by
having a blind spot in the middle of the retina and
corrected by continually moving the eyes.

Page 33

Page 34

CHAPTER XI

RAINBOW SPECTRUM

The center of the visible light spectrum makes up the largest particles. Those to the right and left of visible light are smaller; Those toward ultra-violet and beyond are more dense; Those toward infrared and beyond are less dense (positive +). The more dense particles toward the violet side of the spectrum are more impeded by transparent material than particles toward the red side; Accounting for the Rainbow. The spaces between the color bands is the buildup of bisphere collisions between neighboring colors, creating bands of electromagnetic particles which seem out of place.

Page 35

CHAPTER XII

THE LASER

HOW A LASER WORKS: if you were to remove all but one size of light from a beam these like particles would start to form spheres between them (see planet formation) these spheres would strengthen the force holding the arrangement of particles.   As a laser beam, light particles would force these new spheres into atoms like a lance.   By taking a substance which only allows free parallel motion of one size of light so that other sizes pass out or are smashed and by placing a partial mirror (screen) at either end of the material so that expulsivity is excited and space filled.   Taken that the screen is made up of atoms with a strong negative expulsivity, once the light pushes itself through the curtain of expulsed particles more than the first few are released in a condensed form causing it to expand and wrench apart matter it enters; like freezing water in cracks splitting rock.

www.ingramcontent.com/pod-product-compliance
Lightning Source LLC
Chambersburg PA
CBHW081241170526
45165CB00009B/3149